Bibliographic information published by the German National Library:

The German National Library lists this publication in the National Bibliography; detailed bibliographic data are available on the Internet at http://dnb.dnb.de .

Imprint:

Copyright © 2009 GRIN Verlag, Open Publishing GmbH
Print and binding: Books on Demand GmbH, Norderstedt Germany
ISBN: 9783640870707

This book at GRIN:

http://www.grin.com/en/e-book/148365/the-history-functionality-use-and-advanta-ges-of-the-richter-scale

Enno Eßer

The history, functionality, use and advantages of the Richter Scale

GRIN Publishing

Running head: THE RICHTER SCALE

The history, functionality, use and advantages of the Richter Scale

Enno Esser

Montverde Academy

Abstract

The Richter Scale is a quantitative measure of the magnitude of an earthquake. It has been invented in 1935 by Charles F. Richter and is based the amplitude measured by a seismograph with a correction factor for the distance from the centre of the earthquake from which the measurement is taken. This measurement is totally independent from the damage that it might cause. Although invented in 74 years ago, it is still in use and has many advantages compared to the other scales used for earthquakes.

The Richter Scale

The following paper is about the Richter Scale, a measure used to compare earthquakes to each other. Everybody should have heard about this scale, since it is mentioned in every respectable newspaper. The principle of this scale was invented in 1935 and has not been changed since then. Even though it is a scale for ground motion on the earth, it has been used for measuring thousands of Moon-quakes and two quakes on the planet Mars.

The magnitude of an earthquake on the Richter Scale is calculated from the measured peak amplitude. A peak amplitude is the hight with which the earth goes up. It is measured by several seismographs, spread in the area in which the earthquake occurs.

A very simple example of a seismograph can be found on HowStuffWorks.com and it is simply a large mass hung from the ceiling with a pen, which can just reach a piece of paper on a table under it. If an sufficient strong earthquake occurs, the pen would draw a line on the paper. This may work for one quake, but if there are aftershocks, a measurement over time is needed. This can easily be built in attaching a roll of paper to a motor which slowly pulls the paper over the table.

This simple type of sensor might have been used 100 years ago, but today the sensors are made more sensitive by using levers or electronics and the mass that the seismograph is attached to has a weight of 1,000 pounds or more. The combination of both, levers and heavy weight gives an accuracy, that even gives negative results, which have not been thought of as the Richter Scale was first invented. Nevertheless, these measurements are accepted in the scientific world. To get an idea, which values on the Richter Scale can be cause by which events, a table with various Richter Scale magnitudes and their cause can be found in the table section in the end of this paper.

The further procedure, after the data is collected from one seismograph, is to take the mean value of all the sensors, which took data related to this quake. This is necessary, since the data is often different depending on the underground on which the seismograph is placed and the distance to the centre of the earthquake. The earthquake can be felt differently by the sensor as well, what depends on the nature, the location and the size of the ground motion. Therefore a mean value is evaluated which is plugged in this formula as the hight in mm:

$$M = \log_{10} A\,(mm) + 3\log_{10}[8\,\Delta\,t\,(s)] - 2.92$$

The part of the formula " $+3\log_{10}[8\,\Delta\,t\,(s)] - 2.92$ " is the distance correction factor. It includes the distance between the seismograph and the centre of the earthquake in the calculations. Its functionality is easy, since it works with the difference between the time as the earthquake was created and the time when ground motion reaches the seismograph, which then notices the ground motion. This time is called the S-P time. As an example, if an earthquake occurs at 4.00.00 am and the sensor records the earthquake at 4.00.10 am, the S-P time, Δt, is:

$$4.00.10\text{am} - 4.00.00\text{am} = 00.00.10\text{hours} = 10\text{seconds} = \Delta\,t$$

Richter wrote this system, which includes the distance in the calculations, down in 1958 in his book Elementary Seismology. To estimate an earthquake's magnitude a nomogram has been developed. It solves the mathematical equation quickly and gives a rough estimation of the magnitude.

This nomogram shows a significant attribute as well. An increase of X on the

Richter Scale is equivalent to an 10^x increase of the amplitude. This can be seen as

well.

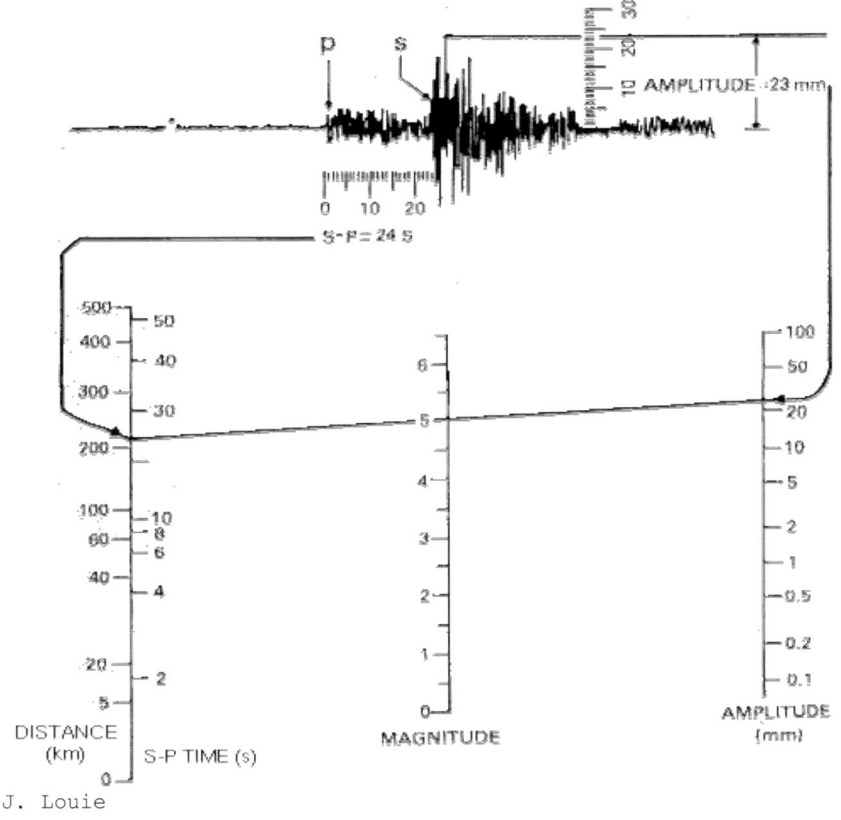

J. Louie

History

Charles F. Richter was born in April 26 in 1900 near Hamilton in Ohio, U.S..

He was an American physicist, who went to the University of Southern California and

continued studying physics Stanford University, where he got his degree in 1920. He transferred to the California Institute of Technology to earned his doctoral degree in 1928. Later on he was employed at the Carnegie Institution of Washington, Pasadena, California and was chosen to work in the Seismological Laboratory until 1936. At this point he changed his career and started teaching at Caltech starting from 1937 to 1970 and combined this with his researches in its Laboratory.

And exactly there, in Caltech, occurred his turning point in his life concerning the Richter Scale, as he met Beno Gutenberg (1889-1960). He was a Caltech professor, born in Germany and a major help to Charles F. Richter as he developed the Scale which is named after him. He also helped him writing a book called Seismicity of the Earth and Associated Phenomena in 1949 and Elementary Seismology in 1958.

As he died in September 30, 1985 in Pasadena, California, he had made himself a name as a great seismological professor and marked his name on the map of the seismological research.

Discussion

The Richter Scale is not the only scale that is used to categorise earthquakes. There are several others categorising earthquakes like the Body wave magnitude, Richter Scale, Moment Magnitude Scale, Surface wave magnitude scale and several intensity scales like the Mercalli Intensity Scale, the Modified Mercalli Intensity Scale, the Japan Meteorological Agency Seismic Intensity Scale, the Medvedev-Sponheuer-Karnik scale, China Seismic Intensity Scale, European Macroseismic Scale.

The differentiation between a magnitude scale and a intensity scale must be made, since they define two different things. All Intensity scales are somehow based on the consequences that an earthquake causes and this is exactly its disadvantage. The consequences can be recorded by asking witnesses and analysing the amount of damage on buildings. The weaker method, the asking of witnesses, is very uncertain, since every person experiences an earthquakes differently what makes it hard to compare two earthquakes.

The better method, the analysing of damage on buildings, has its disadvantage in the amount of time that it takes to look through all the damaged buildings. But again, the damage that is finally recorded is easily falsified, because different designs of buildings, the distance from the epicentre and the type of material the building is made of and build on. Generally spoken, a sandy underground shakes the building, that is build on it, a lot more than a solid underground like rock.

This paragraph is only written about the Mercalli Intensity Scale, but this is the best way to have an idea about the variety of intensity scales. The Mercalli Intensity Scale is simply the oldest scale, that is still in use. For this reason, it is said that all other intensity scales got their idea from the Mercalli Intensity Scale and why all the other intensity scales are somewhat similar to the Mercalli scale. The concept of defining the intensity of an earthquake by analysing its damage can be found in all intensity scales.

All Magnitude scales on the other hand are based on the amplitude collected by seismographs. These scientific instruments are, if calibrated correctly, collect data that is much more able to be compared to each other, what makes the comparing of earthquakes on the Richter Scale much more trustworthy.

Now comparing all these magnitude scales, the Richter magnitude scale is one of the best scales. This is simply because of the fact, that the Richter Scale is understood worldwide, whereas the other scales are only used in specific countries.

Magnitudes in relation to intensities

Transferring magnitudes in intensities is not easy, because an earthquake with a great magnitude can have a small intensity and an earthquake with a small magnitude can have a huge intensity. This is because the intensity depends on the environment where the the earthquake occurs.

For example, an earthquake with a magnitude of 9, the earthquake with the greatest magnitude ever recorded, occurred in Chile in 1960. This earthquake probably did not have an intensity that is as high as an earthquake with a magnitude of 4 in the middle of New York. That is because the intensity is based on the damage that it causes, thus the damage caused in a lonely mountain is nothing compared to the damaged caused in a metropolis. For these reasons, it is hard to compare magnitude scales with intensity scales.

References

Charles F. Richter (2009) In Encyclopaedia *Britannica*. Retrieved April 30, 2009,

from Encyclopaedia Britannica Online:

http://www.britannica.com/EBchecked/topic/502857/Charles-F-Richter

J. Louie (1996)In *What is Richter Magnitude?* Retrieved May 1, 2009,

from Nevada Seismological Laboratory:

http://www.seismo.unr.edu/ftp/pub/louie/class/100/magnitude.html

Michigan Technological University (2007) In *How Are Earthquake Magnitudes*

Measured? Retrieved May 4, 2009, from Michigan Technological University:

http://www.geo.mtu.edu/UPSeis/intensity.html

HowStuffWorks.com (2000) In *How does a seismograph work? What is the*

Richter scale? Retrieved May 7, 2009, from HowStuffWorks.com

http://science.howstuffworks.com/question142.htm

Table

Richter Magnitude	TNT for Seismic Energy Yield	Example (approximate)
-1.5	6 ounces	Breaking a rock on a lab table
1.0	30 pounds	Large Blast at a Construction Site
1.5	320 pounds	
2.0	1 ton	Large Quarry or Mine Blast
2.5	4.6 tons	
3.0	29 tons	
3.5	73 tons	
4.0	1,000 tons	Small Nuclear Weapon
4.5	5,100 tons	Average Tornado (total energy)
5.0	32,000 tons	
5.5	80,000 tons	Little Skull Mountain., NV Quake, 1992
6.0	1 million tons	Double Spring Flat, NV Quake, 1994
6.5	5 million tons	Northridge, CA Quake, 1994
7.0	32 million tons	Hyogo-Ken Nanbu, Japan Quake, 1995; Largest Thermonuclear Weapon
7.5	160 million tons	Landers, CA Quake, 1992
8.0	1 billion tons	San Francisco, CA Quake, 1906
8.5	5 billion tons	Anchorage, AK Quake, 1964
9.0	32 billion tons	Chilean Quake, 1960
10.0	1 trillion tons	(San-Andreas type fault circling Earth)
12.0	160 trillion tons	(Fault Earth in half through centre, OR Earth's daily receipt of solar energy)

(J. Louie)